By the same author:

Between physics and metaphysics: philosophical implications of Quantum Mechanics

First Edition: 2014.
CS Publishing, Seattle 2014.
ISBN-10 : 1505920183
ISBN-13 : 978-1505920185
ASIN : B00RPPHMFO
Copyright © Lucio Giuliodori 2014.

Lucio Giuliodori

*On the Concept of Synchronicity:
Jung between Psychoanalysis and
Quantism.*

Contents

1. What is synchronicity ... p. 12

2. Synchronic experiences... p. 27

3. Synchronicity, the unconscious and quantism... p. 47

4. Conclusion... p. 63

Teaching does not go beyond this limit, that is, of pointing out the way and the journey: but the vision is already a personal work of he who has wanted to contemplate.

Plotinus

1. What is Synchronicity

« The meaning of my existence is that life has addressed a question to me.
Or, conversely, I myself am a question which is addressed to the world, and I must communicate my answer, for otherwise I am dependent upon the world's answer»[1].

This wise Jungian thought encloses the very essence of philosophy: those who are not questioning the world are simply subjected to it; only the understanding of it ensures existential dignity. Jung dared to ask himself critical questions, many of which were thorny, at least for his time, however, it's from this yearning thirst for knowledge, seal of any true philosophy, that scientific knowledge of psychology could advance, freeing itself from

---

[1] C. G. Jung, *Sogni ricordi e riflessioni* (translation of *Memories, Dreams and Reflections*), Milan: Bur 2010, p. 375. My translation.
From now on, all the non-English written books' quotations are translated by me.

the famous sexual neuroses where Freud tended to forcedly place it. Neuroses with which, solely and exclusively, the *whole* psyche was expected to be explained (an attitude that today would be very simplistic, but that was preponderant at Jung's time)[2].

He owns credits to challenge and courage, at the cost of public professional derision that he came up against with his head held high. We know that the Swiss psychiatrist even retired for a long period from the academic world because he was incompatible with it; he kept on doing his research in solitude, deeming it far more important than his academic career

---

[2] Freud was revolutionary and his reflections on sexuality were decidedly original and unusual for his times. Jung, however, was even more so, because in addition to (at first) accepting that vision of the psyche, he was then able to surpass it and by surpassing it, he not only surpassed Freud himself, but also his own personal limits as well.

Freud remained outdated, while Jung, misunderstood and unaccepted both by Freud and by all the academic world, reached unimaginable frontiers, for his time and for ours. He sought the same limits of the psyche that open the doors to a world, that of quantum mechanics, which is still mysterious today: we are not yet able to understand it, nor to describe it. The only attitude that we are allowed to have, for now, is to wonder: the wonder of a sophisticated and elegant, although not yet fully shared, model of understanding reality.

and the profession itself[3]. He was far-sighted, and immediately understood the revolutionary significance of his insights.

Jung went back to his university work in 1933, after twenty years, and in 1952 the essay *Synchronicity: An Acausal Connecting Principle* was released. Here he analyses a phenomenon that immediately stands out for its extraordinary importance, not only for his

---

[3] «During the period when I was working on the images of the unconscious, I took the decision to retire from university, where I had been teaching for eight years, starting in 1905, as a lecturer. [...] Therefore, I consciously gave up my academic career because I could not appear in public before completing my experiment. I felt that something big was going to happen to me and put my hopes in what I thought was more important *sub specie aeternitatis*. I knew it would take my whole life and in order to reach this goal I was willing to take on any risk». C.G. Jung, *Sogni, ricordi e riflessioni*, cit., p. 237.

Understanding how the psyche works, was for Jung one of the most important things there could be, also for the obvious consequences it had on the socio-political context: «Today we can see, like never before, that the danger which threatens us all, does not derive from nature but from man, the soul of the individual and the mass. The real danger is in the psychic aberration of man. Everything depends on whether our psyche functions well or not: if some people lose their heads, today, the result is the launch of the hydrogen bomb». *Ibidem*, p. 171.

study of the psyche, but also for his re-description of reality in its entirety. The scientist immediately realises that the study of synchronicity is a prerequisite to the formulation of a new *Weltanschauung* that necessarily includes some fundamental assumptions of quantum physics.

The study of alchemy, Eastern doctrines, dreams and occult phenomena, led the Swiss psychiatrist to a threshold, at the outset of which the impossibility of mind-matter dualism stands imperious. This obscure threshold is exactly what the concept of synchronicity represents. «Synchronicity is precisely what makes a dialogue between physics and psychology possible, since it involves the entry of subjective elements in physics (external event) and objective elements in psychology (mental state). At this point the universe ends up unfolding in such a manner that the events, both subjective and objective, become an implicit manifestation of the same phenomenon»[4].

> «The object of Jung's reflection is the phenomenon of synchronicity, which

---

[4] M. TEODORANI, *Sincronicità. Il legame tra fisica e psiche da Pauli e Jung a Chopra*, Cesena: Macro Edizioni 2011, p. 82.

according to its definition is the result of two factors: 1) An image that is presented directly (literally) to the unconscious or indirectly (symbolised or suggested to consciousness as a dream, sudden idea, presentiment); 2) An objective matter of fact which coincides with this content. The external event may happen outside the observer's perception, and thus be distant in space, or it may also be distant in time, i.e. it may occur in the future compared to when the psychic event manifested itself to the subject»[5].

Hence, synchronicity is a kind of «temporal coincidence of two or more events that are not connected by a causal relation, and which have the same significant content»[6]. The incredible consequences that this assumption implies are somewhat puzzling - at least for the mechanistic model of interpretation of reality, which was in full force (and largely

---

[5] A. VITOLO, Introduction to C. G JUNG, *La sincronicità come principio di nessi acausali*, Turin: Branded Basic Books 2011, p. 175.
[6] *Ibidem.*

still is) in the mid-twentieth century[7]: «If space and time will be proven to be psychologically related, then also the body in motion must have the corresponding relativity, or must be subjected to it»[8].
Synchronicity was studied in depth by Jung, firstly because he was motivated by that deep and sincere interest in what crosses the phenomena, secondly because not only his patients but he himself experienced several synchronistic phenomena[9], and thirdly

---

[7] We know that Jung was very reluctant to publish his research as he was scared of ending up completely isolated in the academic context, even more than he already was, having rebelled against the authority of Freud, which was at that time indisputable. In addition, as a sign of extreme humility that is an honor for a genius of his caliber, Jung did not feel prepared enough on the subject, therein the first lines of his preface to the work on synchronicity: «By drawing up this paper, so to speak, I keep a promise that for many years I have not dared to fulfill. The difficulties of the problem and its exposure seemed too big; too large the intellectual responsibility, without which such an argument cannot be treated. Moreover, my scientific education seemed too inadequate». *Ivi.* p. 181.

[8] *Ivi*, p. 175.

[9] «In my role of psychiatrist and psychotherapist, I often came into contact with the phenomena in question and, in particular, I have been able to ascertain their importance for the inner experience of man. Most of all it is about things you don't want to

because the Swiss psychiatrist shared this desire for investigation with the Nobel Prize for physics Wolfgang Pauli, his patient and then friend. The two scholars spoke different scientific languages that had to abdicate when facing phenomena that were over and above the same descriptiveness: an evident embarrassment exactly from the scientific point of view - neither physics nor psychoanalysis could tell what certain experiences showed. The two disciplines unveiled a world never seen before, in which what is material is no longer distinct from what is psychic and with regard to this, Pauli says: «We should now proceed to find a neutral or unitary language in which every concept we use is applicable both to the unconscious and to matter, in order to overcome this old belief that the unconscious psyche and the matter are two separate things»[10].

---

say in a loud voice as not to expose yourself to the risk of reckless derision. I have never ceased to be amazed at how many people have had experiences of this kind, and how carefully the inexplicable is guarded. My participation to this problem therefore not only has scientific roots but human ones as well». C. G. JUNG, Preface to *La sincronicità come principio di nessi acausali*, cit. p. 182.

[10] Quoted in M. TEODORANI, cit.

The categories of Western speculative thought (especially those prior to the twentieth century) have a hard time explaining what is non-causal, spaceless and timeless. Synchronicity imposed itself as a «fourth force», a mandala able to harmonise and redefine, clarify and complete the understanding of reality, which is no longer divided and limited: «Space, time and causality, this triad of the classical physical picture of the world, would complete themselves thanks to synchronicity, in a tetrad, i.e. in a *quaternio* that makes an overall judgment possible»[11]. If reality shows phenomena that transcend space, time and causality, this same reality must necessarily be redescribed, most obviously through more integral tools. Jung points out:

> «If these phenomena actually happen, the rationalistic framework of the universe is not valid, because it is incomplete. Thus, the possibility of a reality beyond the phenomenal world, a reality in which other values prevail, becomes a problem from which there is no escape; and we have to take into

---

[11] C. G. JUNG, *La sincronicità come principio di nessi acausali*, cit., p. 274.

account the fact that our world - with time, space and causality - is related to another order of things (which is hidden under or behind it), in which neither the «here and there», nor the «before» and «after» have a meaning»[12].

The study of synchronicity proposes once again the fundamental assumptions of quantum mechanics: the interdependence of observer and observed, non-locality and consequent transcending of space and time categories, review of the mind-body problem and, therefore, «illusion of reality» and «universal harmony». The last two points, respectively, refer to oriental philosophy and Italian Renaissance philosophy, projecting the phenomenon of synchronicity in a purely holistic and indeed *esoteric* context[13].
According to the astrophysicist Massimo Teodorani[14], the reality in which we live is:

---

[12] C. G. JUNG, *Ricordi, sogni e riflessioni*, cit., p. 360.
[13] For this term's specific meaning, at an academic level, please refer to Prof. Wouter Hanegraff's work, *Western Esotericism. A Guide for the Perplexed*. Bloomsbury, London in 2013.
[14] In this essay I will refer several times to Teodorani's studies as the scientist clearly associated Jung to Quantum Mechanics and the latter to all phenomena that transcend mechanistic explanations.

«A reality that can't be defined as neither subjective nor objective. The realms of matter and mind are so inherently interconnected they form a single whole [...]. Yet this concept is not new, but dates back two thousand years when the Hindu Tantric tradition of the world postulated a similar philosophy. According to Tantric philosophy, reality is nothing more than an illusion; this illusion is called «maya». Therefore, the main error that we commit by not perceiving this illusory veil is that we feel separate from the world around us. This is a realm where the laws of classical physics no longer apply. It is about the ultimate goal of physics but also its major stumbling block as it is still impossible to find the metric, the geometric domain and mathematical operators which are able to formally describe it»[15].

Although at that time there was no possibility of contact, Eastern wisdom and Western

---

[15] M. TEODORANI, *Bohm. La fisica dell'infinito*, Cesena: Macro Edizioni, 2006, p. 36.

wisdom had sensed the same truth[16]. According to the Renaissance Philosopher Pico della Mirandola, the world appeared as a *corpus mysticum* of God:

> «Firstly there is the unity of things, thanks to which everything is one with itself, consisting of itself and related to itself. Secondly, it is thanks to this [unit] that a creature is linked to another, and finally all parts of the world form *one* world. The third and main thing is that, by way of it, the whole universe is one with its creator as an army is with its boss»[17].

Quantum physics supports pretty much the same thing, just let us think of the EPR paradox developed by Einstein, Podolski and

---

[16] And this is a distinctive feature of the history of esoteric doctrines: a substantial basic affinity at a conceptual level that goes beyond the barriers of different social contexts and different historical periods. Let us remember the words of Zolla: «the Brahmin practitioner and Platonic teacher are perfectly interchangeable». E. ZOLLA, *Lo stupore infantile*, Milano: Adelphi 1994, p. 37.

[17] Quoted in C. G. JUNG, , cit., p. 252.

Rosen[18], and later validated by Bell's theorem. Bohm himself argues that «the classical idea of a world separated into distinct interacting parts is no longer valid or relevant. Instead, we must consider the universe as an *undivided whole and without fractures*. And thus, we come to an order which is radically different from Galileo's and Newton's: the order of the *undivided whole*»[19].

---

[18] The EPR paradox takes into consideration a simple elementary particle like the electron with no spin. If such a particle is divided into two parts, one must necessarily have spin equal to + ½ and the other spin equal to - ½. This is inevitable to ensure the law of conservation of the spin, of which the sum must be zero - in the moment in which the particles may reconnect. Now, if we launch the two particles at great distances and modify the spin of one of the them, in order to ensure the law of conservation, the other particle must necessarily instantly change its spin. This immediate change however, on the one hand protects the sum of the spins that must be zero, but on the other hand dramatically violates the theory of relativity, which states that a signal cannot travel faster than the speed of light.
[19] D. BOHM, *Universo, mente, materia*, Como: Red edizioni, 1996, p. 176.
In summary, the immediate change of the second particle's spin is really a non-local event, totally unexpected and incomprehensible by classical physics, which cannot explain this phenomenon. In fact, it leads

Aside from mechanism, therefore, the fact that physics and spirit had already met was clear since the beginning of the twentieth century[20]. Synchronistic phenomena only added another afferent piece to the iconic meeting between physics and psychology, a «diptych» that, when able to show itself *completely*, will cause massive and irreversible upheavals, not only in science: «In the case of synchronicity we are not facing a philosophical concept, but an empirical concept that postulates a necessary principle for knowledge»[21]. What characterises synchronic phenomena goes to undermine objective, «material» and everyday reality, erupting and disrupting their macroscopic horizon in which we are the protagonists: it concerns us and sets us in a periplus embodied four-dimensionally.

---

to a fundamental observation: in quantum terms the particles are not separate, they are united.

[20] As a matter of fact it happened much earlier, considering that Young's interference experiment on the dual nature of light (wave and corpuscular) was carried out in 1803.

[21] C. G. JUNG, cit, p. 274.

In my understanding of the world, there exists no physical world as such without the penetration of spiritual and occult energy. And I also think that magic needs not to be explained by physical means, rather what appears as physical to the profane must be explained through magical forces.

Pavel Florenskij

2. Synchronic Experiences

Experiencing co-incidences and synchronicities mean experiencing noumenal gaps, rare moments in which material reality is seriously questioned by an auraticity that cluttering the higher union of psyche and matter, supersedes and displaces the ordinary perceptiveness of reality: the perceiving subject swoops in the perception itself.
In *Auree*, Elémire Zolla bestows a deep qualitative value upon coincidences, though he claims that a more significant value pertains to deeper simultaneity:

> «Just like distant events are scattered in memory, forming eddies in the flow of memories, so it happens in life, where vortexes start spinning flared in a coincidence, in an inexplicable simultaneity, elements that should be separated by time and space. The result is, for those who live those moments, pure wonder: an aura emanates from those overlaps. One is reminded of the metaphor of scholastic philosophy: the angels who are out of the river of time, once in a while immerse one

foot in it. When coincidences happen, it's as if we glimpse an angelic shadow in our world.
To them, considered as arcane manifestations, antiquity bestowed a cult.
You think of a person and he shows up: the moment receives modest consecration, we slip away from ordinary reality, a pure amazement is felt, albeit tenuous, and soon forgotten. A dazzling aura surrounds the more complex simultaneities»[22].

The Jung's studies of synchronistic phenomena, endorsing a non-causal possibility of interaction between psyche and matter, have *de facto* imposed a scientific reassessment of the global description of reality:

«The philosophical principle underlying our conception of the laws of nature is causality. If the relationship between cause and effect shows that it possesses only statistical validity, and only a relative truth, in the final analysis, the causal principle can be applied only relatively to the interpretation of natural processes. As a consequence, it presupposes the existence of one or more

---

[22] E. ZOLLA, *Auree*, Venice: Marsilio 1995, p. 19.

different factors that would be necessary for the explanation of these phenomena. This means that in certain circumstances the link between events differs from causal nature and requires a different principle of interpretation»[23].

Many are the experiences reported by the Swiss psychiatrist that concern both him and his patients. What can be inferred from them, is the role of the unconscious as ally: «The unconscious helps us, as it communicates something to us or produces symbolic allusions. It possesses ulterior tools to inform us of things which with all our logic we could never know. Consider synchronistic phenomena, premonitions, dreams that tell the truth»[24].

In the aim of clarifying this, let me refer to one experience Jung went through. One day, during the Second World War, Jung was returning home from Bollingen by train and wanted to read but could not. As he was about to do so, the mental image of a person that was drowning vehemently assailed him so much as to make him think of the reasons which caused of this image. Well, once home,

---

[23] C. G. JUNG, cit., p. 183.
[24] C. G. JUNG, *Ricordi, sogni e riflessioni*, cit., p. 357.

Jung was informed by his daughter that exactly at the same time he was traveling and had been struck by that thought, his nephew was practically drowning in a pool and was eventually rescued by his older brother. The psychiatrist's deduction was immediately clear: «The unconscious had given me a warning. Why, then, should it not be able to inform me about other things?»[25].

Another synchronicity reported by Jung in his autobiography concerns the premonition of death in a dream. The psychiatrist said he dreamed of his wife who was sleeping in bed, but the latter was gradually transformed into a coffin in which his wife took her last breath and then soared upwards. The dream was so intense that Jung awoke; it was three in the morning. The next day he was told that his wife's cousin had died at the same time as the dream.

Another famous synchronicity is the one of the renowned beetle, and this time there are two witnesses: one of his patients and himself. This patient, during a decisive moment of therapy, dreamed about receiving a golden scarab as a gift and we know the scarab is a symbol of rebirth since the time of the

---

[25] *Ibidem*, p. 358.

Egyptians, especially the dung beetle that feeds its larvae with its stools. Having the ability to turn faeces into nourishment, it is considered a symbol of rebirth and transformation, or, in the language of alchemy, of «transmutation from death into life».

In Jung's office, just as the patient was telling him about the dream, a coleopteran (a similar beetle of the coleopteran order that could be found at such altitudes) appeared fluttering near the window just behind the psychiatrist's shoulders. After this synchronicity, the patient entered a phase of healing, «rebirth», as a matter of fact.

The beetle was considered a symbol of rebirth thousands of years ago by the Egyptians and for this reason, we could infer, it entered the collective unconscious as an archetype. As such, it holds the capacity to act on the psyche of any person, just as each one of us, according to Jungian psychoanalysis, draws on archetypes, from the collective unconscious. (In ancient times, storms were believed to be derived from the wrath of the Gods and humanity seemed to have deeply interiorised such a fear as today many people are still scared of storms)[26].

Thinking of a person and then a moment later seeing him could be said to be a coincidence, Synchronicity instead, presents strong meaningful content which might somehow interact with the person who lives the experience - even at the level of the psyche.

The synchronic phenomenon, as the example of the dream on death shows, can sometimes lead to an experience of pure prescience: the subject sees and describes an event that is happening in the very moment of the description, or that may happen in a more or less near future. As an example, we could mention the famous case of Swedenborg, a Swedish scientist and visionary, who wrote in

---

[26] «After having shown that Freud missed a very important sector of the Unconscious, the one that comes from the feelings that are below the threshold of perception, Jung examines its content. There is, in addition to the personal unconscious, related to past experiences of the individual and explained on those grounds, a collective unconscious that is inherited by the fact that our ancestors have had original experiences too. This is expressed in the unconscious archetypes, which are representative forms - categories - according to which the images revealing indeed the deep collective are formed in us». A. CRESCINI, *Filosofia e psicanalisi*, Brescia: Editrice La Scuola 1993, p. 157.

various books about the transcendent dimensions of space-time that he "visited". It was 1756, and although the mystic was in Gothenburg, he saw the fire of Stockholm and informed all the people near him at that moment by providing accurate details on how the fire was propagating in the capital and then would be extinguished after three hours.

The story of Swedenborg was confirmed with facts. The event is reported by Kant in his book *Dreams of a visionary clarified with dreams of metaphysics*, published in 1766.

Synchrony and foresight, ESP and telekinesis: boundaries, beyond space and time, are quite labile of course and Jung, to his credit, was wholeheartedly fascinated by them, in their entirety:

> «Throughout his life, Carl Gustav Jung was interested in the occult phenomena; often he was even involved, as were his mother and his grandfather too. He thought parapsychology had to be subjected to scientific research, through experiments and theories. Jung was irritated by the fact that the official science of his time, instead of studying and trying to explain them, refused *tout court* occult phenomena. He himself did some experiments with the famous Austrian medium Rudy Schneider and witnessed

psychokinetic phenomena, materialisation, and levitation. He personally experienced a number of spontaneous events: precognition, veridical dreams, telepathic facts, and became deeply interested in alchemy and different kinds of divination such as astrology, tarots, geomancy, and even the ancient Chinese book of oracles *I Ching*»[27].

Among these various paranormal events experienced by Jung, one of the best known is certainly the one regarding Freud's library. If Jung was attracted to the paranormal - and this interest is well documented in the work *Psychology of Occult Phenomena* released in 1902 [28] - Freud was totally uninvolved and definitely sceptical about it - this was just an additional reason for the estrangement between them.

It was 1909 when Jung visited Freud in Vienna in order to express how deeply he was concerned about the urgency of including parapsychological phenomena in scientific research. While Freud was speaking, Jung felt a strange and uncomfortable sensation in the

---

[27] P. GIOVETTI, *I grandi iniziati del nostro tempo*, Roma: Mediterranee 2015, p. 115.
[28] C. G. JUNG, *Psychology and the occult*, New York: Routledge 1987.

diaphragm, which was followed by a loud crash in the library next to them. The sound was so loud that both scientists were frightened. Jung then told Freud that this was a clear example of catalytic exteriorization, though the Austrian psychoanalyst retorted that it was all nonsense. Jung then told Freud not only that he was wrong, but also there would be another exteriorization following that one. In fact, that is what happened: Jung had not even finished speaking when from that same library came another crash which was similar to the first. Freud's distrust grew even more and the relationship between them became even more difficult.

In addition to Jung and Swedenborg, among the various personalities belonging to the history of ideas who have experienced synchronicity, we can't omit to mention Pauli, or rather «the Pauli effect».

The Nobel Prize winner in Physics in fact had the ability to trigger psychokinetic events wherever he would go. This was dangerous especially in laboratories where his only presence could cause all the instruments to go haywire. In fact, he was banned from entering several laboratories of experimental physics and this specific fact overtly reveals the actual power of the Pauli effect itself.

Pauli turned to Jung, with whom he underwent an intense and fruitful therapy, which ended up laying the foundations not only for his healing but also for a solid friendship. The physicist had problems in his private life; we could infer that the emotional sphere of his personality was repressed by the intellectual side, and the Pauli effect phenomena were signs that the unconscious was sending to the conscious in order to attract attention on his emotional and unconscious side, which was particularly active but was constantly lured by the intellectual and rational side[29].

The therapy, as well as the healing, also favoured that wonderful exchange of ideas that saw the two scholars agree on the urgency of the recognition of a new science that had just been born before their eyes: psychophysics[30], coryphaeus of noetic

---

[29] Several scholars through their research have insisted on the idea that diseases are nothing more than «messages» that the unconscious launches to the conscious. To this regard, see the work of Joe Dispenza, *Evolve Your Brain*, Delfield beach: Health communications 2007.

[30] «The sessions with Jung not only helped Pauli to recover from his depression, but also created a partnership that would bring the two thinkers to lay the foundations of the «physics of consciousness» in

horizons put together in sync. The missing piece of its scientific legitimacy depended only on the limits of language; Jung and Pauli had taken a road that opened wonderful and unexpected scenarios which could even cross the borders of Quantum physics itself, i.e. a new science able to incorporate the material into the psyche and vice versa:

> «Pauli himself, who in virtue of the great genius he was gifted with and the well-deserved Nobel Prize in physics, was a thinker who knew how to look well beyond his nose, at least much more than his bigot colleagues of the clerical entourage of "positivist scientism". In fact, he had no difficulty saying: *Neither the language of physics - the first - nor the language of psychology - the second - are effective enough. In reality, the unconscious speaks a language that is physic-symbolic (third language), which we have to transform into a «neutral language» (the fourth language), which can be understood by the rational consciousness. To my mind, managing to find this fourth language, a neutral language, will be the starting challenge of the twenty-first century*»[31].

---

order to try to understand the problem of synchronicity in particular». M. TEODORANI, *Sincronicità*, cit., p. 42.
[31] *Ibidem*, p. 87.

The major problem for the two thinkers did not concern the alleged reality of synchronic phenomena, already largely taken for granted, but the human ability to comprehensively describe them in linguistic terms, with a view to determine a final scientific admissibility. The source of synchronicity's reliability is not repetition - we cannot produce these phenomena by command since we are not yet able to control the unconscious that is the place they come from - but the «simple» reality of their (albeit sporadic and indescribable)[32] manifestation. Jung said in no uncertain terms that only one phenomenon of synchronicity is enough to decree existence, and therefore scientific relevance: if it happens, it should be studied, «the unconscious is real because it acts», as he argued in his work *The Psychology of the Unconscious*[33], and the way it acts is a synchronous mode. Synchronicity is part of the world in which we live, it is part of us, it

---

[32] The question of infrequency is however relative, as Jung advocates that an awareness of the phenomenon increases its manifestation: understanding and internalizing the synchronicity produces and increases synchronic phenomena.
[33] See. C. G. JUNG, *The Psychology of the Unconscious*, New York: Routledge 1946.

concerns us in any case, it pertains to us *a priori*.

In addition to usual synchronicities there are also multiple synchronicities. The following example clarifies their definition. In 1948, Pauli was invited to preside over the opening of the Institute of Psychic Sciences, wanted and created by Jung. As soon as he entered the Institute, a Chinese vase full of water fell on the floor - without having been moved by any apparent cause - spilling all the water it contained. The lavish symbolism underlying the event refers primarily to the bond with the East (Chinese vase) and its philosophy, widely appreciated and revived by Jung. Secondly it refers to the word "flood" (which in English means the overflow of water) very similar to the surname of the famous alchemist Robert Fludd (the importance of Jung's studies on alchemy as the basis of his theory of the unconscious are rather renowned)[34]. Lastly, there is an evident analogy with water itself, a symbol of purification.

In the major arcana tarot cards, the card of Temperance presents an angelic figure that is holding two vases from which a stream of

---

[34] See: C. G. JUNG, *Psychology and Alchemy*, New York: Routledge 1980.

water is flowing. This card is number 14 and it comes after the card of Death: the skeleton of the latter is thus transmuted into an angel.

Temperance is also a figure of equilibrium, in fact even if the figure is winged, it firmly remains stuck on the ground, engaged in a practical and concrete work: the flow of water in and out of the two pots, symbolises energy in circulation that is purified by harmonising opposites which do not even exist anymore, absorbed by the reconciler flow.

This can't but refer to the same Pauli, purified by the Jungian therapy and back up and running to his scientific industriousness, strengthened by a spiritual gift: wings, i.e. the ability to reconcile the opposing rationality and emotion in his life. The broken vase also recalls the astrological symbol of Aquarius, which in turn refers to the new age of Aquarius, an era that scholars of astrology - and Jung being one – indicate as a massive renovation at the level of consciousness, purified and enhanced by a more powerful scientific, and at the same time "spiritual", knowledge[35]. Furthermore, the astrological

---

[35] With regard to Jung's specific and explicit interests in astrology, refer to the second chapter of the already mentioned essay on Synchronicity, entitled *An Astrological Experiment*, where the scholar proposes a

symbol of Aquarius is exactly a man who is holding a jar from which water pours out. It is the man who spills the water. It's easy to spot Pauli himself in this figure (or his unconscious that «gives the command» to break the vase) who - referring back to the above mentioned opposites reunited by the Temperance - attempts to merge the old physics (Mechanism) with the new (Quantism that supports the synchronicity and the non-locality): «Pauli, having studied in depth and then compared with each other Kepler's ternary thinking (space, time, causality) and the great alchemist Robert Fludd's quaternary thought (space-time, psyche, causality, synchronicity) , felt he belonged to both of these schools of thought, and in fact had no difficulty in stating *I bear Kepler as much as Fludd in me, and for me this is a necessity in order to reach, as best as I can, a synthesis of this pair of opposites*»[36].

According to Jung, if you are able to read and interpret reality through its symbols, as in the

---

detailed and deep study of astrological symbolism and meaning underlying it, which are relevant in terms of acausal and synchronistic phenomena.

[36] M. TEODORANI, *Sincronicità*, cit. p. 89.

case of a dream (or a work of art)[37], not only do you increase the release of additional symbols and synchronicity, but you can also understand it in its occult, symbolic and *a priori* dimension: unconscious.

---

[37] It's not a case that alchemy is called *Ars Regia*: to live alchemically means to live creating a work of art, the *Grande Opera* in fact, the one relative to a further dimension of being human, a human that is transcended, transmuted, deified. Mircea Eliade, author of the basic text *Arti del Metallo e Alchimia*, claims that the true originality of alchemy lies in the premise from which it moves: the conception of a complex and dramatic life of Matter. Alchemy treats matter as Deity is treated in the course of the Mysteries: as a God, matter suffers, dies, and is reborn. For Eliade, there is a parallelism between the path that leads to the transmutation of minerals and the process by which the alchemist comes to his moral and spiritual rebirth. The adept works on the substance to be purified and to awaken himself, to get hold of the divine power which sleeps in his being. In the work quoted above, the great scholar cites a phrase from Oswald Croll, a disciple of Paracelsus, which gives us the measure of the perfect synchronisation between the task of the Great Work and the perfections reached by the adept: «Alchemists are holy men, who by virtue of their deified spirit have tasted the first fruits of Resurrection in this life and have had the opportunity to foretaste the heavenly kingdom». G. GANGI, *Misteri esoterici. La Tradizione ermetico-esoterica in Occidente*, Roma: Mediterranee 2006, p. 239.

Consciousness is the theater, and precisely the only theater on which everything that takes place in the Universe is represented, the vessel that contains everything, absolutely everything, and outside which nothing exists.

Erwin Schrödinger

3. The Unconscious, Synchronicity, Quantum Physics.

«If in the quantum realm Wolfgang Pauli discovered that the ultimate laws of nature are not subject to the principle of causality - they are nothing more than a mandala of shapes that synchronize matter and interconnect it in all its parts – in the psychological and cognitive realm Carl Jung revealed out the existence of the collective unconscious, intended as an objective reality and a base substrate, the purpose of which is to synchronously join together both psyche and matter»[38].

The obvious implications with quantum mechanics inhere in the principal Jungian concepts of *personal unconscious* and *collective unconscious*[39]. According to Jung when we

---

[38] M. TEODORANI, *Sincronicità*, cit., p. 75.
[39] The first was for Jung a kind of receptacle that exists and *acts* beyond our control: it is a big part of us of which we are unaware. We are not aware of it, it is indeed unconscious, that is it lives and works beyond the level of our attention.

forget something, in reality we do not forget it at all, it comes out of the sphere of our conscious to enter into the unconscious[40]. «Its energy charge is so diminished that it can no longer manifest in the conscious, but, even if it has been lost for the conscious, it is not lost for the unconscious»[41]. As evidence of this, the Swiss psychiatrist relates the fitting example of reading:

«Suppose that there are two people one of which has never read a book, while the other

---

[40] Although it is difficult to determine «where», since the concepts related to location exclude themselves *a priori* when we speak of the unconscious. Probably more than within us, memories accumulate in an area beyond space and time in which we, however, have access by virtue of the non-local reality in which we live. «Non-local» does not mean that it is not «located», rather that it works transcending every possible location, even though it may *seem* objective. However, it does appear to be objective, because we do not possess descriptive tools that are appropriate to a non-mechanistic and non-dualistic model: a synchronic model. This leads to Rupert Sheldrake's famous morphic fields and Karl Pribram's holographic brain, as we shall see shortly.

[41] C. G. JUNG, *The Psychology of the Unconscious*, cit., p. 141. Zolla himself says: «So in a meditative and curious mind the many readings do not fall inert and stacked, but they connect orchestrating, creating little by little: literature». E. ZOLLA, *Auree*, cit., p. 14.

has read a thousand. Suppress from the mind of both readers all memory of the past ten years, during which the first was simply living, while the second was reading a thousand books. In this instance both will be equally ignorant, but anyone will be able to tell which of the two has read the books and, most importantly, has understood them. The experience of reading even if forgotten for a long time, leaves behind traces where you may find the vestiges of the past. This durable indirect influence depends on the fixation of the impressions that are all preserved, even when they are no longer able to emerge into consciousness»[42].

So this is the territory of the personal unconscious, to which Jung adds the well-known collective unconscious, as in the realm of the individual's unconscious there are also ideas that do not belong to his private history: «What kind of ideas are they? In short they are mythological fantasies, which do not correspond to any event or experience of the personal life of the individual, but only to myths»[43].

---

[43] C. G. JUNG, *The Psychology of the Unconscious*, cit., p. 141

Jung's deduction is obviously the following: «If these ideas do not spring from the personal unconscious, i.e. from the experiences of individual life, then where do they come from?»[44] Therefore, the collective unconscious acts as an atavistic coryphaeus of all those enmeshed potentials which, if reacquired, efficaciously extend the awareness of the individual, who in this way could find innate faculties of creative definition of reality. In order to explain this, Jung mentions the Kantian categories, even though we know that the difference is substantial. In fact the Kantian categories are meant to organise and construct the phenomenal world by providing a physical knowledge of scientific nature. On the other hand, for the Swiss psychiatrist it is the noumenal world, using Kantian language, to be affected by the categories of his psychoanalysis. Jung himself even uses the term «noumenon» as synonymous with «unconscious»[45].

---

[44] *Ibidem*.

[45] «The collective unconscious should be understood as an extension of man beyond himself as a rebirth into a new dimension, that dimension which was literally represented in the ancient mysteries. [...] On no account it should be read as something related to inherited ideas. However, there are possibilities of innate ideas, a priori conditions for the production of certain patterns,

Pauli himself, in some ways «was quite sure that that invisible matrix able to hold the world together was the collective unconscious, to which the personal unconscious occasionally accessed through dreams full of meaning and through synchronicity»[46].

In a nutshell: «Jung knew that the unconscious is not found in a known space, but in a kind of hyper-spaced dimension with its laws clearly differentiated from those of causality known to standard science. The synchronism between the mental state of an individual and an event in the world of matter showed even too well that, in addition to the known laws of physics, there are others which we still do not know well»[47].

This hyper-spatial dimension belongs to the collective unconscious, a sort of Platonic hyperuranion where there are «ideas of things» or, precisely in Jungian terms, the archetypes of things[48]. We are all connected

---

in a sense akin to Kantian categories. Although these innate conditions do not produce in themselves any content, yet they give a definite shape to the content already acquired». M. TEODORANI, cit., p. 142

[46] M. TEODORANI, cit. p. 64.

[47] *Ivi*, p. 21

[48] Interesting is the analogy with the Akasha proposed by Teodorani: «Real scientific discoveries are born first

to such *a priori* of reality and the way we connect to it lies in the symbol. Dreams, synchronicities and all paranormal phenomena fall into what Jung described as a symbolic communication that allows the subject to interact with the collective unconscious.

«Just as the individual is absolutely not a unique being and separate from others, but is also a social being, so the human psyche is not a phenomenon closed in itself and purely individual, but is also a collective phenomenon. [...] The universal similarity of human brains implies the universal possibility of uniform psychic functioning. This functionality is the collective psyche»[49].

---

as intuition of a higher reality. The only way to access it is to connect to the realm of archetypes, which is nothing more than a huge library containing all the knowledge in a symbolic universe. In the end this realm beyond time and space and the mysterious and mythical Akasha, which has been handed down in the Eastern traditions, are exactly the same thing». *Ibidem*, p. 69.

[49] C. G. JUNG, *The Psychology of the Unconscious*, cit.,, pp. 110-111.

Unconscious, memory, brain, and collective psyche... In the light of the discoveries of modern physics, everything really does seem connected. The neurosurgeon Karl Pribram has endorsed the Bohmian theory of the holographic nature of reality by dint of numerous studies with rats, to which a part of the brain was removed. Despite several subsequent removals, the rats continued to preserve memories, with regard to which, therefore, following the results of the experiments, it was impossible to admit a localized existence. The same human ability to draw instantly on any memory, between billions and billions of pieces of information can but confirm the non-localization of memories, and therefore the non-classifiable nature of time.

According to Pribram's theory of the holographic brain, memories are in a dimension outside of the brain to which, however, the latter seems to have access.

Even the theory of morphic fields of the philosopher and biologist Rupert Sheldrake supports this thesis. After that a group of monkeys, who lived on a Japanese island, had acquired the ability to wash sweet potatoes before eating them, it was discovered that earlier, another group of monkeys living in another island had acquired the same technique. The two groups were obviously not in physical contact but the information, according to the theory of morphic fields, had travelled non-locally and synchronously reaching other members of the same species.

Several years ago, in a television interview that is still visible on YouTube, Franco Battiato was asked what he thought of those ascetics who retire from the world to live in solitude and meditation, an attitude that at first seems to lead to an escape from the world more than a commitment in favour of an improvement. The famous singer told the interviewer that one could not even imagine how good those people did for humanity with their attitude of «apparent» escape. With the theory of morphic fields in fact the Battiato's thesis is easily explainable and understandable: if the ascetic achieves a high degree of spiritual development, through morphic resonance, synchronously the other members of his species would be affected - even if unconsciously of course - as the change goes to affect the unconscious rather than the conscious[50]. "Changing the world by

---

[50] Therefore an alternative theory of evolution, whose themes are triggered by the whole Western esoteric tradition, let us think of Steiner and Theosophy but also

changing ourselves", a well-known esoteric truth seem to echo here [51].

Coming back to the topic of memory, Bruno's techniques themselves were already intended to promote an improvement of reality beginning with the individual, the initiate:

«Bruno does not only strengthen the mnemonic muscle, he wants to change the cosmos. That is - let me say again - changing the very structure of the mind of every initiate and then, through it, the world, through the interdependence of micro and macrocosm, that is mind-universe. This is the work of

---

Gurdjieff or the Italians Evola and Assagioli, all engaged in the definition of a system, both practical and speculative, able to set in motion an improvement to the «biological machine», in the Armenian master's words.

As we mentioned previously, the same alchemical Great Work is nothing but the story of an improvement: transmuting the mortal into the immortal, making it divine in a process artfully accomplished: *Ars Regia* indeed.

[51] It's not surprise that the various esoteric traditions focused on the evolution of the individual rather than focusing on the community, the concept of «Absolute Individual» in Evola is particularly illustrative to this regard. The fierce zollian criticism to the «mass-man» (uomo-massa) is also indicative, in contrast to the qualitative model of the initiate.

"his" hermeticism: change man's mind, changing his intimate essence and then, though him, a new born man, revolutionize the world»[52].

As microcosm and macrocosm are indissolubly united, man himself is both the subject and the object of the same reality: if he changes, synchronously reality changes.
In quantum mechanics the observer changes the observed. In fact if we carry out in laboratory a measurement of the trajectory of a very fine particle such as a photon, an electron or an atom, our measurement changes the position of the particle, i.e. the measurement does not presuppose a deterministic procedure. In the context of the infinitely small, the mechanistic model is blatantly contradicted. This is why in quantum mechanics we speak of «probabilism», as all we can do is to consider a range of probability as to the position of the particle. In this regard a special mathematical function was created to define the possible position of the particle: the wave function.

---

[52] G. LA PORTA, *Giordano Bruno. Vita e avventure di un pericoloso maestro del pensiero*, Bologna: Bompiani 2001, p. 195.

It is called wave function because between one possible measurement and another, the particle dissolves in a superposition of probability waves and it is therefore potentially present simultaneously in a number of different sites; only the act of measurement makes the particle collapse in a given location. This absurd behaviour of particles has been shown in the famous paradox of Schroedinger's cat locked in the box: inside the box, the cat is both alive and dead, and only the act of opening (measure) determines the actual life or death of the cat, which is «in the hands» of the observer.

As the physicist Pascual Jordan says: «Not only do the observations disturb what should be measured, but they produce it... We force an electron to take a definite position ... But we ourselves produce the results of measurement»[53] . In very simple words: we are involved in the measurement that is not only a «measurement» but it is an act of modification of the real. This means above all that we cannot even see

---

[53] Quoted in M. TEODORANI, *Entanglement*, Macroedizioni, Cesena 2007, p. 9.

objective reality for what it is, we can only see our own interpretation of it, since we are editing it.

Therefore, quantum entanglement confirms the interconnection of reality at a deep level and it seems to represent a clear frame of meaning to the phenomenon of synchronicity, because if we do translate the same concept of interconnection to macroscopic reality (although we still cannot prove it in mathematical terms) we can easily deduce that it reproduces the same phenomenon experienced in micro reality: «it is also important to keep in mind that if all we know about entanglement concerns the behaviour of elementary particles, then nothing prevents us from thinking that the same can happen when you put elementary particles together to form living organisms»[54]. In fact Teodorani, going further in depth, asserts that:

«The process of life is intimately connected with the process of the observer who looks at reality. A particle is probably not able to consciously observe reality. Definitely a particle that interacts with reality changes it, but does not do it consciously. On the other hand, a bio-system's own psyche is able to

---

[54] *Ibid*, p. 105.

interact as an «observer» with the observed in a fully conscious way. The reason why Jung's mysterious phenomena of synchronicity happen only to us, sophisticated biological beings, is no coincidence»[55].

Becoming aware of this intrinsic power will probably be part of the many challenges that twenty-first century science is going to face. However, what Jung has produced is exactly a breaking of the levees between science and mysticism, between archetype and object, between the conscious and the unconscious.
Hence, science must also recognize itself as part of this wider spectrum of knowledge. a knowledge which includes man first of all. This science should consequently re-evaluate man's inner world as the *a priori* of a phenomenal objectivity and as its (magic) mirror.

---

[55] *Ibid*, p. 130.

Space is the metaphysical aura of the environment. The environment is the spiritual projection of human actions.

Enrico Prampolini

4. Conclusion: towards new cognitive horizons.

Taking everything into account, it should be argued that if synchronicities happen - and they do objectively happen - the Newtonian mechanistic paradigm, by all means, must be called into question: the primary paradigm in charge from Galileo to the present day, is no longer acceptable. As far as science itself is concerned, that method can no longer deemed valid for an inclusive explanation of reality as it simply is not exhaustive. In so far as that method can't integrate all those phenomena that transcend reality and challenge it in the laboratory, that method not only is unproductive but it also intrudes upon an objective and reliable explanation of the world we live in.

If the various esoteric traditions of all time didn't but implicitly challenged and discredited causality and determinism, today it is just physics that challenges and discredits itself by imposing a narrative that will not only affect science but also the psyche *in primis* – something never faced before.

The ruthlessly shocking extent of the new scientific paradigm, that creeps between the shaky boundaries of rationality, brings about cognitive dissonances which need to be seriously assessed by the contemporary scientific community. The latter will hold the responsibility to draw a brand new description of the world in which we live in, bestowing it a new deeper uncanny meaning which could floor us but with which we cannot fail to grapple with. In a nutshell, they should eventually acknowledge that reality is definitely something different, different in its very depths, the picture painted by Classical Mechanics doesn't apply with what Synchronicity posited, it can't even be likened to it, so deep the difference is.

A sort of noetic revolution set off, new truths have been unveiled, different cultural and scientific horizons can be glimpsed, it's up to us pondering upon them, brimming the gaps with the (by now) unenforceable old *Weltanschauung*. We are called to veer towards this new, puzzling body of symbolic ideas, ensuring that its coherent – yet indescribable - set of concepts could eventually give rise to a crucial breakthrough, backbone of a compelling relationship with what surrounds us and fills us within.

There are no two separate worlds, two different realities, a normal world and a paranormal world... There is a single world, that you can "look at" or "see".

Carlos Castaneda

REFERENCES

D. BOHM, *Universo, mente, materia*, Como: Red edizioni 1996.

A. CRESCINI, *Filosofia e psicanalisi*, Brescia: Editrice La Scuola 1993.

J. Dispenza, *Evolve Your Brain*, Delfield beach: Health communications 2007.

G. GANGI, *Misteri esoterici. La Tradizione ermetico-esoterica in Occidente*, Roma: Mediterranee 2006.

P. GIOVETTI, *I grandi iniziati del nostro tempo*, Roma: Mediterranee 2015.

C. G. JUNG, *The Psychology of the Unconscious*, New York: Routledge 1946.

C. G. JUNG, *Psychology and Alchemy*, New York: Routledge 1980.

C. G. JUNG, *The Psychology of the Unconscious*, Rome: Newton Compton 1989.

C. G. JUNG, *La sincronicità come principio di nessi acausali*, Turin: Branded Basic Books 2011.

C. G. Jung, *Sogni ricordi e riflessioni*, Milan: Bur 2010.

G. LA PORTA, *Giordano Bruno. Vita e avventure di un pericoloso maestro del pensiero*, Bologna: Bompiani 2001.

M. TEODORANI, *Sincronicità. Il legame tra fisica e psiche da Pauli e Jung a Chopra*, Cesena: Macro Edizioni 2011.

M. TEODORANI, *Bohm. La fisica dell'infinito*, Cesena: Macro Edizioni, 2006.

M. TEODORANI, *Entanglement*, Cesena: Macroedizioni 2007.

E. ZOLLA, *Auree*, Venice: Marsilio 1995.

www.luciogiuliodori.net

www.luciogiuliodori.net

www.luciogiuliodori.net

Made in United States
Troutdale, OR
05/28/2024